BOOK I.

Relativity: An Expanded View of Gravity and General Theory

by Frederick Taouil

I. Scientific Method & Science	1
II. The Visionaries	3
III. Everything is Relative - Follow the Light	7
IV. Einstein - General Relativity	13
V. Einstein's Space-time and Causality	16
VI. A Quantum Leap - The Birth of Quantum Physics	17
VII. Quantum Gravity & Time	20
VIII. Quantum Theory's Standard Model	23
IX. Higgs Boson - A New Truth Emerges	25
X. Relativity: An Expanded View of Gravity and General Theory	28
XI. Relative Gravity: Motion of Planet Mercury	34
XII. Last Word is Unifying: What's Next?	45

Frederick Taouil

I. Scientific Method & Science

The scientific method has its merits.

Throughout history, the methods for confirming theories and ideas has been done primarily through a structured approach of observing and noting evidence, which proves or denies the validity of predictions and models given by those theories. This approach is taught and ingrained from an early age and continues throughout all academics as an integral part of science. Yes, the scientific method is a necessary function of science because it separates fact from fiction through the evidence of empirical data.

Too often, our scientific community proclaims this process of observing as the their key to understanding the truth of science without true virtue.

It is unfortunate, but science has grown accustomed to overstating the contribution of observers without giving proper credit to the visionaries who help us see farther into the looking glass so we can understand more of "the truth."

Without the proper challenges in place to press the scientific community to advance its knowledge of the universe, academics and scientists have developed a overarching tone of complacency and stagnation in all but a few fields of physics. Science seems content

Book I. Relativity

with simply trying to understand what has already been given as "science".

This lack scientific fomentation has succeeded in limiting innovation and stymied any real creativity. Where is the imagination?

The purpose of this book is to help break the mold and stir up the pot of an anemic community; to give imagination a new pedestal to preach from and to give science a new lens to explore new waters in an uncharted realm of possibilities and understanding. With each new discovery, we set ourselves free from the restraints that shackle our minds.

After reading this book, I want you to believe that there is so much more to learn and to understand. I want you to see the laws of physics for what they are: mathematical models and ideas given to help predict and explain the universe around us; they are not truth in itself.

In other words, to know that three gallons of gas will provide enough fuel for a car to drive 60 miles does not mean that you understand how a car works.

There is so much more to discover, so let us begin.

II. The Visionaries

A clear understanding of the knowledge and ideas given by our predecessors is fundamental to presenting new ideas on the same subject. This understanding can be qualified and quantified by a person's ability to interpret those ideas into their own words. Redefining an idea proves some form of understanding; but being able to expand on those ideas shows complete understanding. There are not many people who have this ability and vision, but those who do, have inevitably reshaped how those around us observe the world.

Bruce Lee was one of those extraordinary people who had the understanding and vision to create his own form of martial arts. At an early age, Bruce Lee mastered a style of kung fu known as *Wing*

Chun. After being seriously challenged in 1967, Lee realized the shortcomings of *Wing Chun* and worked to create a more fluid, less structured form of combat. Combining several fighting styles, Lee evolved a unique style of martial arts called *Jeet Kune Do or the Way of the Intercepting Fist*, which he made famous throughout the martial arts world, and through film.

Leonardo Da Vinci was also a man of incredible vision. His curiosity was insatiable. His understanding of the world around him gave him incredible insight. He was a man with a 'feverishly inventive imagination'. His insight literally helped him to draw, paint, and sculpt a world never seen before. He sketched the first parachute, first helicopter, first aeroplane, first tank, first repeating rifle, swinging bridge, paddle boat and first motor car. He shared his 'vision' through the extraordinary talent and skills of his mind and hands. His magnificent works of art continue to reign and inspire today in the glory of his accomplishments during the Age of Renaissance.

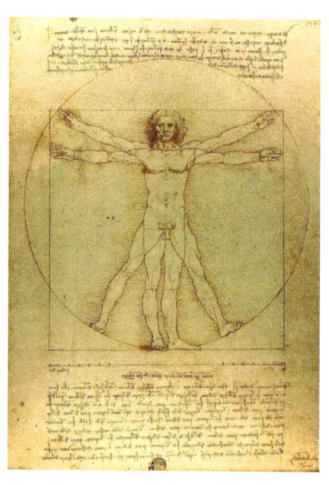

What makes men like Bruce Lee, and Leonardo Da Vinci visionaries? What qualities separate them from those who merely follow their lead and call forth science or art in their names? It is their ability to understand the world around them and successfully interpret the world in new ways.

Book I. Relativity

In the age of modern science, Albert Einstein is considered one of its best-known visionaries. His ability to understand empirical data and ideas presented to him in theoretical physics helped him discover more of the truth about our universe. Einstein embodies and exemplifies all the characteristics of a true visionary. He has been quoted as saying his most amazing work was a result of "seeing pictures" in his head and envisioning the answers before arriving at his scientific postulations. It was only through tapping into his imagination and vision that he was able to explore new truths about the universe and give us new ways to predict the world around us.

Are there other truths ready to be found?

Yes, there are. We only need to know where to look.

> *To myself I seem to have been only like a boy playing on the seashore, and diverting myself in now and then finding a smoother pebble or a prettier shell than ordinary, whilst the great ocean of truth lay all undiscovered before me.*
> *Isaac Newton*

III. Everything is Relative - Follow the Light

In 1868, the equations of James Maxwell, a Scottish mathematician and physicist, explained that all electromagnetic waves traveled at exactly the same speed as light in empty space. Maxwell concluded with his experiments that light and other electromagnetic waves travel at a fixed speed. That speed, a constant, has become a cornerstone of physics denoted in scientific notation as c.

Book I. Relativity

Light Speed[1] c = 299,792 km/s

By chance, but furthering their research on the properties of light, the Michelson - Morley experiments performed in 1887 unexpectedly demonstrated that light travels at a constant speed - regardless of whether measured in the direction of the Earth's motion, or at right angles to it. In other words, the speed of light does not increase or decrease with the motion of the source.

The speed of light is *absolute*.

For the first time in physics the relative characteristics of a body in motion were discovered to be the same as a body at rest when viewed from the inertial frame at rest. The speed of light never changes. This simple observation has lead to the development of some of the most powerful concepts of the 20th century.

What do we mean when we use the term *relative*? It describes a "frame of reference" to which motion and rest may be measured. Thus, any set of points or objects that are at rest relative to one another enables us to describe the relative motions of bodies. This dynamic portrayal of motion then leads to the idea of an "inertial

---

[1] In a vacuum

system," or a reference frame relative to which motions have distinguished dynamical properties[2].

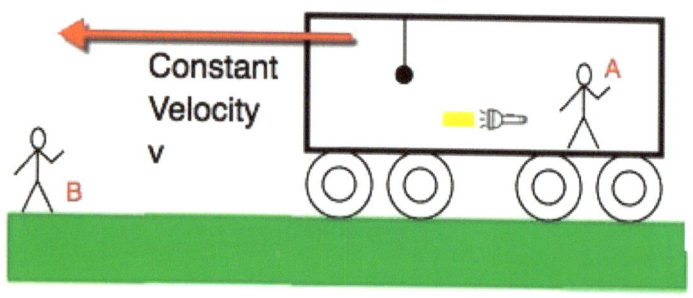

Figure 1. Frames of reference

Let's illustrate.

Imagine a man standing in a railroad car on a moving train traveling at a constant speed, labeled in *frame A*. Although the car is moving on the train, the frame of reference is the car itself *(A)* and the mechanical laws of physics apply in the car as if the man were standing still.

Next, imagine a man watching the railroad car pass by him, shown in *frame B*. The frame of reference for that man would be the world around him *(B)*, which is not moving. The mechanical laws of physics apply to the man in *frame B* as they apply with *frame A*.

---

[2] These properties will be described by Einstein as relativistic properties, which we will opine on.

Book I. Relativity

When the man in *frame B* describes the man in *frame A*, he's speaking about how *frame A* moves *relative* to his stance in *frame B*.

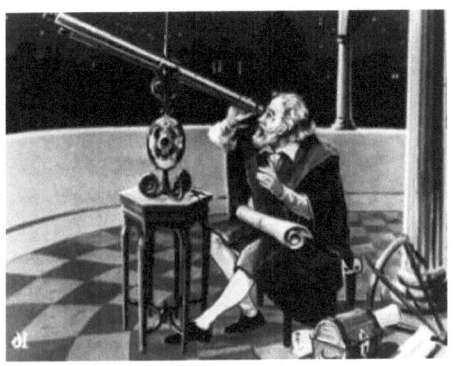

Galileo Galilei

The concept of relativity is not new.

The great Italian physicist best known for his work in astronomy, Galileo Galilei (1564-1642), argued that the mechanical laws of physics are the same for every 'inertial observer,' or relative observer. He explained that objects moving uniformly at a constant speed in a straight line also follow the same laws of physics as a body at rest. His work became known as the 'principle of relativity' or 'principle of invariance.'

Late in the 19th century, the recently discovered *absolute* of the speed of light *(c)* posed a problem with concept of relativity for physicists.

Let's describe the problem by returning to our illustration of the two men in *frame A* and *frame B (Figure 1, pg 9)*. If the man in *frame A* shines a flashlight at the man in *frame B* while the railroad car is traveling towards him at velocity *v*, classical physics says the man in *frame B* should see light traveling toward him at *x* speed which is equal to the speed of light *(c)* plus the velocity of the railroad car *(v)* or $x = c + v$. But in fact, since *c* is an *absolute*, the speed of light is still *c* $(x = c)$ regardless of what the frame the man is in.

How can this be?

In 1905, Einstein solved the riddle with his theory of "special relativity." He described objects in motion (the man in *frame B observing frame A*), as having *relativistic* properties different from properties described by the classical laws of physics. Those relativistic properties included an effect called "time dilation", where time was mutable and not as absolute as was originally thought.

Einstein's theory on *time dilation* explained that in order to keep the system of *frame A and B* in equilibrium, as the man in *frame B* observes the flashlight beaming light from the moving railcar in *frame A* at the constant speed of *c*, time must change. To the observer in *frame A*, time is relative and consistent in his inertial frame (clocks will act normally). But when *frame A* is observed by an observer in *frame B*, the observer in *frame B* will "see" time moving slower in *frame A* to keep *frame A* in equilibrium with *frame B*. This concept is both fascinating, and counter-intuitive.

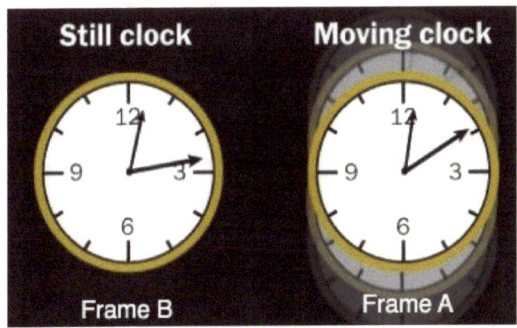

As a result of his observations, Einstein discovered that space and time were intrinsically infused into a single continuum that he called *space-time*. Observers of different frames of reference could have different experiences with time depending on a frames relativistic motion to an inertial frame.

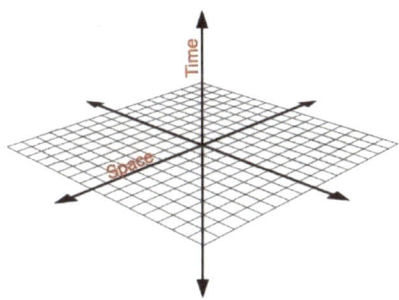

Special relativity combines space with time, into a concept called space-time.

IV. Einstein - General Relativity

Einstein was not done.

For the next ten years, he grappled with ways to formulate the interactions of massive objects into his space-time model.

Let's pause, and point out Einstein's extraordinary talent as a visionary. Einstein said he developed part of his theory of general relativity, the theory of equivalence, by "envisioning" a man in freefall on earth, while enclosed in an elevator. That man would float freely inside the chamber, just as he would if he were sitting in deep space where there is no gravity. He then compared that to a woman accelerating in outer space in the same elevator where that woman would be pulled to the floor just as she would be pulled by gravity on earth. He noted that in each example, the man and woman would not know if they were in space or on earth. From such equivalent actions of acceleration, Einstein postulated his equivalence principle.

Book I. Relativity

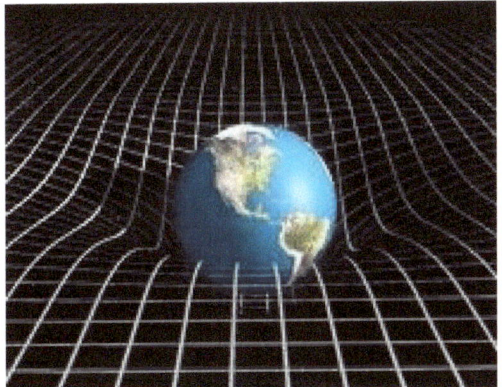

During this period, Einstein realized that gravity was not a force at all, but an effect created by the warping of space-time itself. He concluded that the curvature of space-time as it directly related to matter caused the observed effects of gravity. This relationship is specified by Einstein's field equations, which are a system of partial differential equations. In 1915, Einstein detailed his findings with his published theory of general relativity.

To be clear, Einstein's theory of general relativity (1915) generalizes his earlier work on special relativity (1905) and Newton's law of universal gravitation, giving us a unified model of gravity as a geometric property of space-time.

Einstein described space-time as a place that could become warped with valleys and ridges. Think of a bowling ball on a trampoline where a massive object (the bowling ball) interacts with space-time (the trampoline) causes gravity (the deformation of the trampoline).

This space-time curvature is caused by the energy–momentum of matter. Paraphrasing the relativist John Archibald Wheeler (1911-2008), space-time tells matter how to move and matter tells space-time how to curve.

V. Einstein's Space-time and Causality

In physics, an effect cannot occur *before* its cause. In Einstein's relativity, causality means that an effect cannot occur from a cause that is not in the past. Furthermore, a cause cannot have an effect in the future. These observations are consistent with the grounded belief that causal influences cannot travel faster than the speed of light and/or backward in time. In the world of quantum field theory, regarding observable events with a spacelike relationship, "elsewhere", has to commute, so that the order of observations or measurements of such observables do *not* impact each other.

What is important here is that in Einstein's relativity, there is a clear cause, followed by an effect. In quantum theory, a cause and effect can happen at the same time so long as they are happening in different places. In both concepts, cause never happens before an effect.

VI. A Quantum Leap - The Birth of Quantum Physics

At the turn of the 20th century, scientists believed that they understood the most fundamental principles of the universe. Atoms were the solid building blocks of the universe. People trusted the Newtonian laws of motion. Most of the problems of physics seemed to be solved. However, starting with Einstein's theory of relativity which replaced Newtonian mechanics, scientists realized that their knowledge was far from complete. Of particular interest was the growing field of quantum mechanics, which completely altered the fundamental tenets of physics.

Max Planck is considered the father of the quantum theory.

As Einstein was working on his precepts of relativity, the German physicist Max Planck published a groundbreaking study of the effect of radiation on a "blackbody" substance in 1900. His work has become the starting point for quantum theory in modern physics. Planck's study introduced what was to prove a revolutionary concept in physics. He verified the theory that the oscillators comprising the black body and re-emitting the radiant energy incident upon them could not absorb this energy continuously, but only in discrete amounts, or a *quanta* of energy.

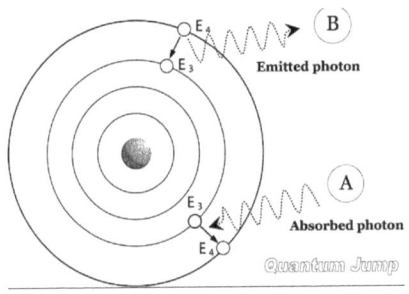

In the illustration above, when energy is added to the system electrons "jump" to another sphere of orbit. This "jump" is a measure of quanta.

This concept of energy quanta fundamentally conflicted with all past physical theory, and its importance was not fully appreciated at first. However, the evidence for its validity gradually became overwhelming as its application accounted for many of the discrepancies between observed phenomena and classical laws of physics. In 1918 Planck's contribution was recognized when he was awarded the Nobel Prize in Physics "for the discovery of energy quanta."

Through his experiments, Planck demonstrated energy exhibits characteristics of physical matter in certain situations. In contrast to theories of classical physics, energy is solely a continuous wave-like phenomenon, independent of the characteristics of physical matter. Planck's theory held that radiant energy is made up of particle-like components, known as "quantum." The theory helped to resolve previously unexplained natural phenomena, such as the behavior of heat in solids and the nature of light absorption on an atomic level.

Other scientists, including Albert Einstein, Niels Bohr, Louis de Broglie, Erwin Schrodinger, and Paul M. Dirac, advanced Planck's theory and made possible the development of quantum mechanics–a mathematical application of quantum theory that maintains energy is both matter and a wave, depending on variables. Quantum mechanics thus takes a probabilistic view of nature, sharply contrasting with classical mechanics, in which all precise properties of objects could be calculated.

Today, the combination of quantum mechanics along with Einstein's theory of relativity is the basis of modern physics.

VII. Quantum Gravity & Time

Quantum physics, quantum gravity, and the quantum theory of gravity refers to attempts to combine quantum mechanics and general relativity into a functional unified theory. The unified theory tries to describe the force of gravity according to the principles of quantum mechanics and is an overall effort towards the holy grail of physics, a "theory of everything;" a theoretical theory describing the whole truth of our universe. Quantum theory and general relativity, while coexisting nicely in most respects, appear to be fundamentally incompatible at unapproachable events such as singularities of black holes and the Big Bang itself. Many scientists believe that a combination of the two theories is essential to acquiring a real handle on the fundamental nature of time itself.

Any theory of quantum gravity must deal with the "problem of time." Time has a different meaning in quantum theory than general relativity.

In quantum mechanics, time is universal and absolute; its steady ticks dictate the actions between particles. However, in general relativity, time is relative and dynamic within a dimension that's inextricably interwoven into x, y and z, giving a us the four-dimensional model of "space-time".

To add to the perplexing nature of the "problem of time", other views given by physicists show time to not exist at all. In the famous Wheeler-DeWitt note published in 1967, the two physicists showed mathematically how time disappeared completely from their equations - thus suggesting at its most fundamental level the universe is timeless. In reaction to the Wheeler-DeWitt equation, some have concluded that time is a type of fictitious variable in physics; that we are possibly confusing the measurement of different physical variables with their actual existences. This is comparable to what we thought was the existence of the "force" of gravity, which is not a force at all, but an effect to space-time warping as described by Einstein.

More recently, Stephen Hawking has added a new twist on the idea of time, introducing the concept he describes as imaginary time. Although rather difficult to understand, imaginary time is not imaginary in the sense of being unreal or made-up, rather, it has similar characteristics to mathematics and the imaginary number scale. It can perhaps best be portrayed as an axis running next to that of regular time. It provides a way of looking at the dimension of time as if it were a dimension of space, so that it is possible to move forward and backward along it, just as one can move right and left or up and down in space. This idea of imaginary time effectively ties Einstein's general relativity with quantum theory models.

VIII. Quantum Theory's Standard Model

All particles and their subatomic interactions can be described almost entirely by a quantum field theory called the Standard Model. The Standard Model consist of 61 elementary particles. Those elementary particles can combine to form composite particles, accounting for the hundreds of other species of particles since the models introduction in the 1960's.

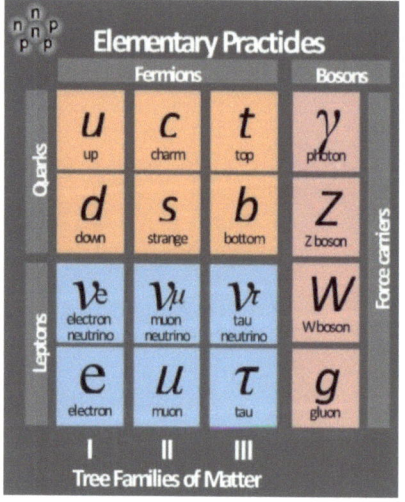

Although the Standard Model has been self-consistent and demonstrated much success with providing experimental predictions, it leaves some phenomena unexplained. The model does not incorporate the full theory of gravitation as described by general relativity, and does not account for the accelerating expansion of the universe as possibly described by dark energy. The model does not contain any viable dark matter particle that possesses all of the required properties deduced from current observational cosmology.

However, for all that it does not give, the Standard Model is an amazing evolution of particle physics stemming from classical notions of atoms, protons, neutrons, and electrons.

Although the word "particle" can refer to various types of very small objects, "particle physics" usually investigates the smallest detectable particles and fundamental interactions necessary to explain their behaviors. With our current understanding, elementary particles are excitations of quantum fields that also govern their interactions. Modern particle physics generally investigates the Standard Model and its various possible extensions, to the newest "known" particle, the Higgs boson, and as well, the oldest known force field, gravity.

IX. Higgs Boson - A New Truth Emerges

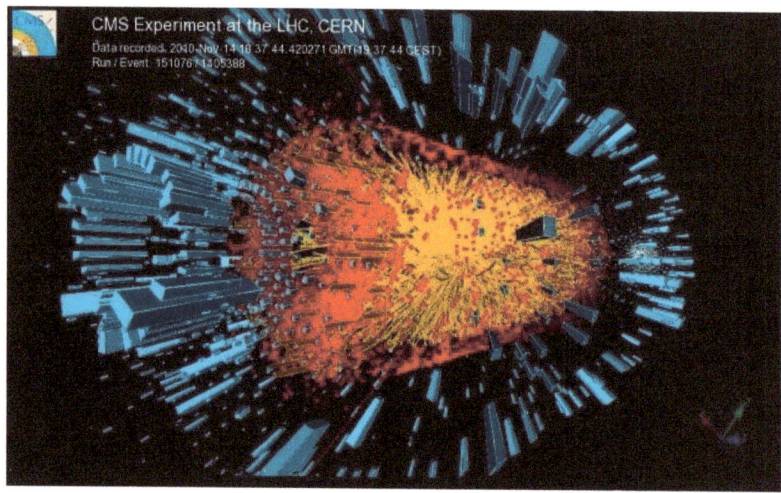

Let's now delve into the Standard Model and into one of its most recent and inspiring observations made at the *European Organization for Nuclear Research* (CERN) Large Hadron Collider, located in Switzerland. The Higgs boson, sometimes referred to as the 'God Particle', was finally confirmed to exist on July 4th, 2012 at CERN after its first description in the early 1960's. Its observation has helped solidified the Standard Model and has given credence to the existence of the Higgs field. The importance of this cannot be overstated.

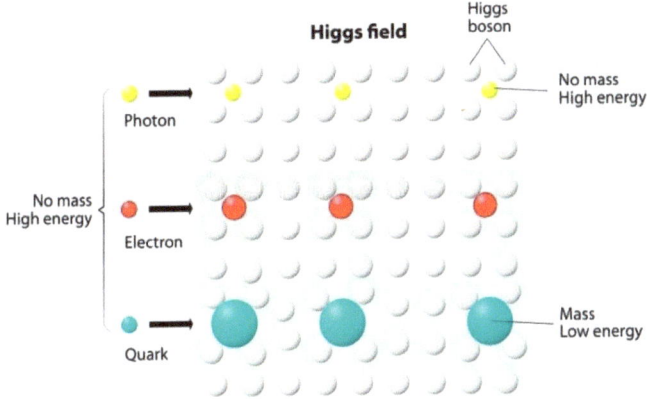

The Higgs field is an energy field existing everywhere in the universe. As particles pass through the Higgs field, they interact with the field and are "given" mass, similar to an object passing through molasses. Particles then become slower than the speed of light because of this mass.

Some particles, such as photons (light), pass through the field unobstructed by the field and travel without added mass at the speed of light, expressed as c. In contrast, other particles such as quarks interact with the field and that interaction causes mass.

It's important to understand that mass itself is not generated by the Higgs field. Mass is "given" to particles from the Higgs field via the Higgs boson, which contain relative mass in the form of energy. Once the field has endowed a formerly massless particle with mass, the particle slows down because it has become heavier.

In other words, if the Higgs field did not exist, particles would not have the mass required to 'attract' one another and would just float around freely at light speed. Gravity, too, would not exist.

The process of giving a particle mass is known as the "Higgs effect." This effect transfers mass or energy to any particle that passes through it. Because light is a wave, it passes through the field and gains energy, not mass.

Book I. Relativity

X. Relativity: An Expanded View of Gravity and General Theory

What if gravity were an effect caused by particles interacting with the Higgs field?

Not only are particles 'given' mass by the Higgs field as proven in 2012 at CERN, but the particles interactions with the field also create the effect we know as gravity.

Boat traveling through water

Let's illustrate this idea through an analogy.

Imagine a boat traveling through water. As the boat floats, it displaces water to 'give' itself buoyancy; we can draw a parallel to a particle which is 'given' mass in the Higgs field. As the boat travels through the water, it also creates a wake behind it; this wake is analogous to gravity. Particles 'moving' through the Higgs field create a 'wake of gravity'.

This illustration is important as it shows how mass and gravity are infused into a relationship that is both intuitive and powerful. You can visualize how gravity and mass are part of the same expression of a particle moving through the Higgs field, yet are different in nature.

Let's add a second boat of exactly the same size into the water. Both boats displace exactly the same amount of water, previously described as mass. The mass of both boats are equal. Now let's make one boat travel twice as fast through the water as the other. Although the mass of each boat is exactly the same, the faster boat will create a much larger wake than the smaller boat.

You can visualize that the speed of the boats directly affect the size of the wake even when the boats are exactly the same.

This wake effect is the same effect happening at the subatomic level where particles interact with the Higgs field. The effect of a particle moving 'faster' through this field creates more gravity than the particle of the same mass which moves slower. We will describe the gravity attributed to the faster particle when measured against the slower particle as *relativistic gravity*.

Book I. Relativity

We are not changing Einstein's theory of general relativity, rather, we are expanding upon it. The motion of massive bodies in our universe are well explained by Einstein's models, but the true meaning of this motion is not.

What general relativity does not explain completely is how gravity, like time, is relativistic; a body at rest has different gravitational properties than a body in motion when observed by the *inertial observer*.

The motion of particles causes them to interact with the Higgs field which creates gravity; the more motion a particle has, the more gravity the particle will have. With this precept, we'll extend Einstein's mass-energy equivalence equation to include gravity and show how it is governed by a particles relativistic motion.

It is important to understand that when we say *motion*, we mean both the motion through space[3] and the motion through time. We discussed earlier how Einstein's concept of space-time combines space and time. It is the same when we describe motion. If an object is at rest in an inertial state, it will still be in motion because it is moving through time.

To add, two object traveling at the same speed can be described as having no relative motion with regards to *relativistic gravity*; the two frames are equal, and the interactions with the Higgs field by both

---

[3] measure by speed

object would be the same. Motion in our hypothesis is given by the difference of an object's speed relative to another, which can be described as *relativistic* in nature.

The Expanded Law of Gravity

Gravity is an effect of particles interacting with the Higgs field. The effects of gravity follows the principles of general relativity when observed from an inertial frame. Additionally, a body in 'motion'[4] will have the same gravitational effects as a body at rest. But the effects of gravity, like time, are relativistic in nature and will change with motion when observed from the inertial frame. A system's relativistic properties can be calculated by using Einstein's mass-energy equivalence equation, $E=mc^2$.

Einstein's beautifully simple equation, $E=mc^2$, tells us that there is a direct relationship between mass and energy measured by motion. It states that both mass and energy are interchangeable. This interchangeability can be directly measured by the velocity of a mass in the system, given by a factor of c.

In general relativity, *relativistic* mass is different than relative mass; relative mass is the mass of a body at rest, while *relativistic* mass is the mass of a body in motion. *Relativistic* mass is measured by

---

[4] motion without acceleration

converting the energy of a mass in motion using E=mc², back to a mass equivalent as if the body were at rest.

We will see that the *relativistic* mass of the body has a direct relationship with *relativistic gravity*. It will be our premise that *relativistic gravity* is already given by today's observed gravitational forces as recorded by institutions like the National Aeronautics and Space Agency (NASA). What we aren't given today is the *relative gravity* or relative mass which would be observed if one were in the 'inertial frame' of a body in motion as if we changed our frame to observe as an inertial observer.

We now have a new lense to view the universe and can use this to calculate a never before understood *relative gravity*.

The formula for *relative gravity* can be represented as.:

$$g_{rel} = g_{earth}(v_{rel}/v_{earth})(\sim c)$$

Where *g* is gravity, *v* is velocity and ~c is a gravitational variable.

We use earth as the inertial frame of reference for gravity and mass because every observation made about the mechanics of motion of celestial bodies has been made from earth's 'frame of reference'. We will correct this error with our new understanding and expand on the meaning to *relativity* in general.

Where can we find our expanded laws of gravity in action in the universe?

The more relativistic motion(velocity) a mass has, the more we'll be able to see the effects of *relativistic gravity* in action and the larger the mass is the better.

There is no better place to look than the planets in our solar system. Furthermore, the mysteries and anomalies which come with the planet Mercury give us the opportunity to explain that this planets extraordinary characteristics are, in fact, not that extraordinary.

Book I. Relativity

XI. Relative Gravity: Motion of Planet Mercury

First, it is important to note that the density and mass of our planets and sun are still given by equations that are centuries old. They are not derived from sophisticated scales or by eccentric scientific methods, but rely on an old equation from the 1600's, called Newton's law of universal gravitation. The below exemplifies how mass is still equated today:

"Knowing the mass and radius of the a planet and the distance of the planet from the sun, we can calculate the mass of the sun (*right*), again by using the law of universal gravitation. The gravitational attraction between the planet and the sun is G times the sun's mass times the planet's mass, divided by the distance between the planet and the sun squared. This attraction must be equal to the centripetal force needed to keep the planet in its orbit around the sun. The centripetal force is the planet's mass times the square of its speed divided by its distance from the sun. By astronomically determining the distance to the sun, we can calculate the planet's speed around the sun and hence the sun's mass."

Scientific American

Data calculated from these equations are used in NASA's current planetary fact sheet, which describes the characteristics of planets, including their gravity, mass, and density.

NASA's Planetary Fact sheet

	MERCURY	VENUS	EARTH	MOON	MARS	JUPITER	SATURN	URANUS	NEPTUNE	PLUTO
Mass (10^{24} kg)	0.330	4.87	5.97	0.073	0.642	1898	568	86.8	102	0.0146
Diameter (km)	4879	12,104	12,756	3475	6792	142,984	120,536	51,118	49,528	2370
Density (kg/m^3)	5427	5243	5514	3340	3933	1326	687	1271	1638	2095
Gravity (m/s^2)	3.7	8.9	9.8	1.6	3.7	23.1	9.0	8.7	11.0	0.7
Escape Velocity (km/s)	4.3	10.4	11.2	2.4	5.0	59.5	35.5	21.3	23.5	1.3
Rotation Period (hours)	1407.6	-5832.5	23.9	655.7	24.6	9.9	10.7	-17.2	16.1	-153.3
Length of Day (hours)	4222.6	2802.0	24.0	708.7	24.7	9.9	10.7	17.2	16.1	153.3
Distance from Sun (10^6 km)	57.9	108.2	149.6	0.384*	227.9	778.6	1433.5	2872.5	4495.1	5906.4
Perihelion (10^6 km)	46.0	107.5	147.1	0.363*	206.6	740.5	1352.6	2741.3	4444.5	4436.8
Aphelion (10^6 km)	69.8	108.9	152.1	0.406*	249.2	816.6	1514.5	3003.6	4545.7	7375.9
Orbital Period (days)	88.0	224.7	365.2	27.3	687.0	4331	10,747	30,589	59,800	90,560
Orbital Velocity (km/s)	47.4	35.0	29.8	1.0	24.1	13.1	9.7	6.8	5.4	4.7
Orbital Inclination (degrees)	7.0	3.4	0.0	5.1	1.9	1.3	2.5	0.8	1.8	17.2
Orbital Eccentricity	0.205	0.007	0.017	0.055	0.094	0.049	0.057	0.046	0.011	0.244
Obliquity to Orbit (degrees)	0.034	177.4	23.4	6.7	25.2	3.1	26.7	97.8	28.3	122.5
Mean Temperature (C)	167	464	15	-20	-65	-110	-140	-195	-200	-225
Surface Pressure (bars)	0	92	1	0	0.01	Unknown*	Unknown*	Unknown*	Unknown*	0.00001
Number of Moons	0	0	1	0	2	67	62	27	14	5
Ring System?	No	No	No	No	No	Yes	Yes	Yes	Yes	No
Global Magnetic Field?	Yes	No	Yes	No	No	Yes	Yes	Yes	Yes	Unknown
	MERCURY	VENUS	EARTH	MOON	MARS	JUPITER	SATURN	URANUS	NEPTUNE	PLUTO

Book I. Relativity

Mercury, the planet closest to our sun, travels 1.6 times faster than the earth around the sun. Curiously enough, Mercury's mass and density are abnormally high when compared to other celestial bodies of the same size and volume.

There are many articles published about the abnormal mass of Mercury. Here is one from Space.com:

"Mercury is so dense, scientists believe its heavy iron core accounts for two-thirds of the planet's mass, more than twice the ratio of core to mass for Earth, Venus or Mars. Scientists aren't sure what caused this incredibly high density, but suggest it might have started off with more mass that got scraped off by collisions. Researchers hope MESSENGER's geology measurements can shed light on how the planet formed, and how it got to be so dense."

Space.com

Mercury is perfect for us to examine because it holds all the facets we'd expect our hypothesis on *relativistic gravity* to correct; Mercury's seemingly extraordinary mass given by today's science books is much more ordinary than our scientists think.

With our *Expanded Law of Gravity* we know that Mercury's particles have increased interactions with the Higgs field when compared to Earth's particles because Mercury travels 1.6 times faster than Earth. This increased interaction directly causes a larger effect of gravity we refer to as relativistic gravity. Thus, when we are observing Mercury, we are seeing a skewed version of relative gravity which is changed due to the Mercury's relativistic motion[5]. The relative gravity of Mercury can be calculated by removing Mercury's motion from the equation.

What's Mercury's relative gravity?

NASA's fact sheet notes that Mercury's gravity as 3.711 m/s² (Earth's gravity is 9.807 m/s²).

Planet Mercury Facts
Orbital Speed: 47.4 km/s (1.6x Earths)
Density: 5.43g/cm³(.98x Earths)
Gravity: 3.711 m/s²

---

[5] Note that if the speed of Mercury were the same as the speed of Earth then both would be in the same relative space-time and you could say relativistic gravity = relative gravity.

Using our expanded law of gravity, the overall gravity or *relativistic gravity* of a Mercury *(z)* is equal to the relative gravity *(x)*, plus the gravity given by the mass in 'motion' *(y)*. Earth is used as the inertial frame of reference so earth's velocity around the sun will be the basis for calculations. The mass in motion is the measured velocity greater than Earth's velocity. Since Mercury's velocity is 1.6 times greater than Earth's, this faster velocity adds gravity to the overall system *(z)*. The equation can be express as:

$$x + y = z$$

- *x*, relative gravity
- *y*, gravity given 'motion' relative to earth
- *z*, *relativistic gravity*

where *relativistic gravity (z)* is the current gravity observed in today science books.

Substituting the current gravity given on NASA's fact sheet in our equation, we'll have the following:

$$x + y = 3.711 \ m/s^2$$

This equation is important as it illustrates relative motion *(y)* having an additive effect to a system's overall gravity.

We can solve for *x*, relative gravity, by making an assumption that we can apply the direct inverse relationship of Mercury's velocity as compared to Earth's in the form of a factor, which we'll call *relative factor*.

We'll then use this *relative factor* to help express density in a new "relative" way.

Relative factor, f can be written as:

$$f = v_{earth} / v_{Mercury}$$

Remember, Earth is the inertial frame.

Since,

$$\text{Earth's orbital velocity} = 29.78 \text{ km/s}$$
$$\text{Mercury's orbital velocity} = 47.4 \text{ km/s}$$

We solve for *f*

$$f = 29.78 / 47.4$$

or

$$*f = 0.625$$

Book I. Relativity

*Note that this is simply and inverse of the speed of Mercury as it relates to Earth and should only be taken as a loose interpretation for generalization of the expanded theory of gravity and general relativity.

Applying the relative factor to the *relativistic gravity* we solve for relative gravity. :

$$x = f(z)$$

or

$$x = .625 \, (3.711 \text{ m/s}^2)$$

or

$$x = 2.319 \text{ m/s}^2$$

As expected, the relative gravity on Mercury is less that the *relativistic gravity* as viewed on Earth.

We can continue applying this factor to get the relative density and mass of Mercury.

Let's solve for relative density as we can use this to make some further observation. Mercury's density given on NASA's fact sheet is:

$$\text{Mercury's density} = 5.51 \text{g/cm3}$$

Applying the *relative factor* we'll get,

$$\text{relative density} = 5.51 \text{g/cm3} \, (0.625)$$

or

$$= 3.44375 \text{g/cm3}$$

Logical Deduction

Here comes the fun part. What celestial body in our solar system is comparable to Mercury in size?

It so happens we don't need to look very far beyond our own moon. The moon is two-thirds the diameter of Mercury's.

Moon   Mercury

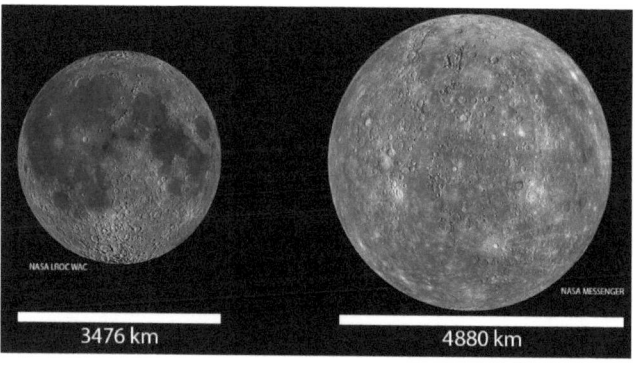

Book I. Relativity

What works well with using the moon as a comparison is, we can also assume the relativistic velocity of the moon is the same as earth's, thereby negating the need for a *relative factor* which had we applied to planet Mercury earlier.

So, what is the density of the Moon?

We can use Newton's law of universal gravitation here and simply give what is stated in NASA's fact sheet:

$$\text{Moon density} = 3.34 \text{ g/cm}^3$$

Compare this to our new calculated density of Mercury.

$$\text{Mercury's relative density}^{6} = 3.44375 \text{g/cm}^3,$$

Perfect! You can you see the application of the *Expanded Law of Gravity* to solve for Mercury's relative gravity and 'real' density holds merit and seems incredibly logical, ordinary, and plausible.

Mercury's density is not so abnormal after all.

The application of *relativistic gravity* and the idea of *relative factor* becomes even more plausible when we understand that in planetary

---

[6] Calculated by using the *Expanded Law of Gravity*

science there is a standard process for building celestial bodies roaming our heavens. The process is known as planetary differentiation. It basically describes that planets of the same size and volume form in very similar ways and should not deviate much from their "common" counterparts.

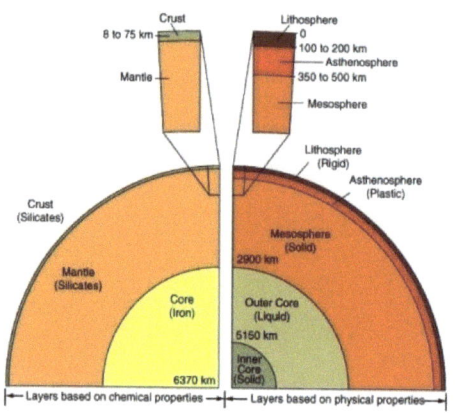

A pictorial explanation of what constitutes normality with celestial object formations of bodies as small as the Moon and as large as the Earth.

Let's clarify "common." Celestial bodies that are less than 3 times the size when measured by their diameter are considered common in planetary differentiation. Jupiter and Earth are not common with their characteristics since Jupiter, a gas giant, is 11 times the diameter of planet Earth, a terrestrial planet.

When comparing Mercury's current mass and density, recorded on NASA's fact sheet, to other "common" celestial bodies in our solar system, and applying the science of planetary differentiation, the numbers simply *do not* adds up. And since science does not have an explanation for Mercury's abnormal characteristics, it's simply written off as an anomaly. Mercury's high density of mass and formation as a planet is said to have happened from "an extraordinary occurrence of nature."

With the *Expanded Law of Gravity* giving us Mercury's relative density and mass, we now have a *new* truth.

Frederick Taouil

XII. Last Word is Unifying: What's Next?

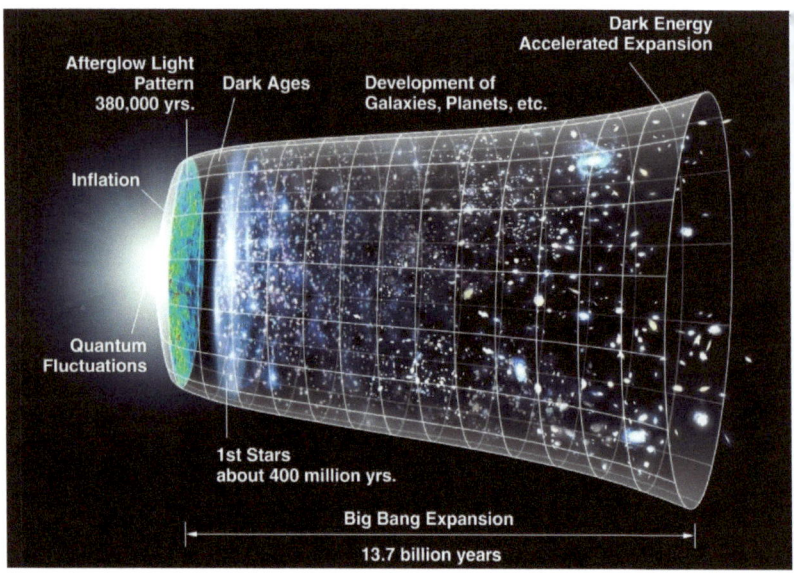

The Big Bang theory is the widely accepted cosmological model for the universe from its earliest existence to the present. The model describes how the universe was created from a very high density and high-temperature state. Furthermore, it offers a constructive explanation for phenomena, including the abundance of light elements, the cosmic microwave background (CMB), an evolution of stars and galaxies, and as well, Hubble's law. Big Bang theory surmises that the beginning of our universe came from a singularity of space-time.

Book I. Relativity

In the beginning there was nothing, but what was created was everything.

Space and time did not exist.

The creation of space and time from nothing can be represented mathematically as zero.

In physics, the law of conservation of energy states that total energy of a system remains constant - it is always conserved.

Zero is nihility.

Therefore nihility must be conserved in some way. Nihility is the natural state of everything.

With the creation of space-time by the Big Bang, how would the laws of conservation be applied? Would the creation of space-time have an opposite effect conforming to the laws of physics?

Imagine a balloon being blown up with air creating space within the balloon giving the balloon it's 'existence' or state of being. What is the natural state of the balloon? The tension of the walls of the balloon want to bring the balloon back to its natural state. The energy given by air counters this natural state.

Could gravity be analogous to the walls of the balloon trying to bring the universe back to its natural state? Could this understanding help us piece together the 'theory of everything'?

"I have not particular talent. I'm merely inquisitive." - Albert Einstein

www.ingramcontent.com/pod-product-compliance
Lightning Source LLC
Chambersburg PA
CBHW040245220526
45473CB00001B/378